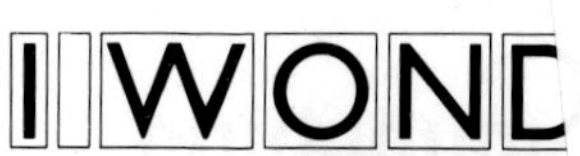

FARM ANIMALS

Jean Watson

Pictures by Ann Blockley

I wonder why chickens
have voices that squawk,
It isn't a very polite way
to talk.

I wonder why cats
have pink tongues that lick,
I suppose they are useful
for keeping fur slick.

I wonder why dogs
have legs that can bound,
They're certainly useful
for leaping around.

wonder why pigs
have noses that snuffle,
A styful of snufflers
makes quite a kerfuffle.

I wonder why horses
have horseshoes that clatter,
Instead of soft paws
that would go pitter-patter.

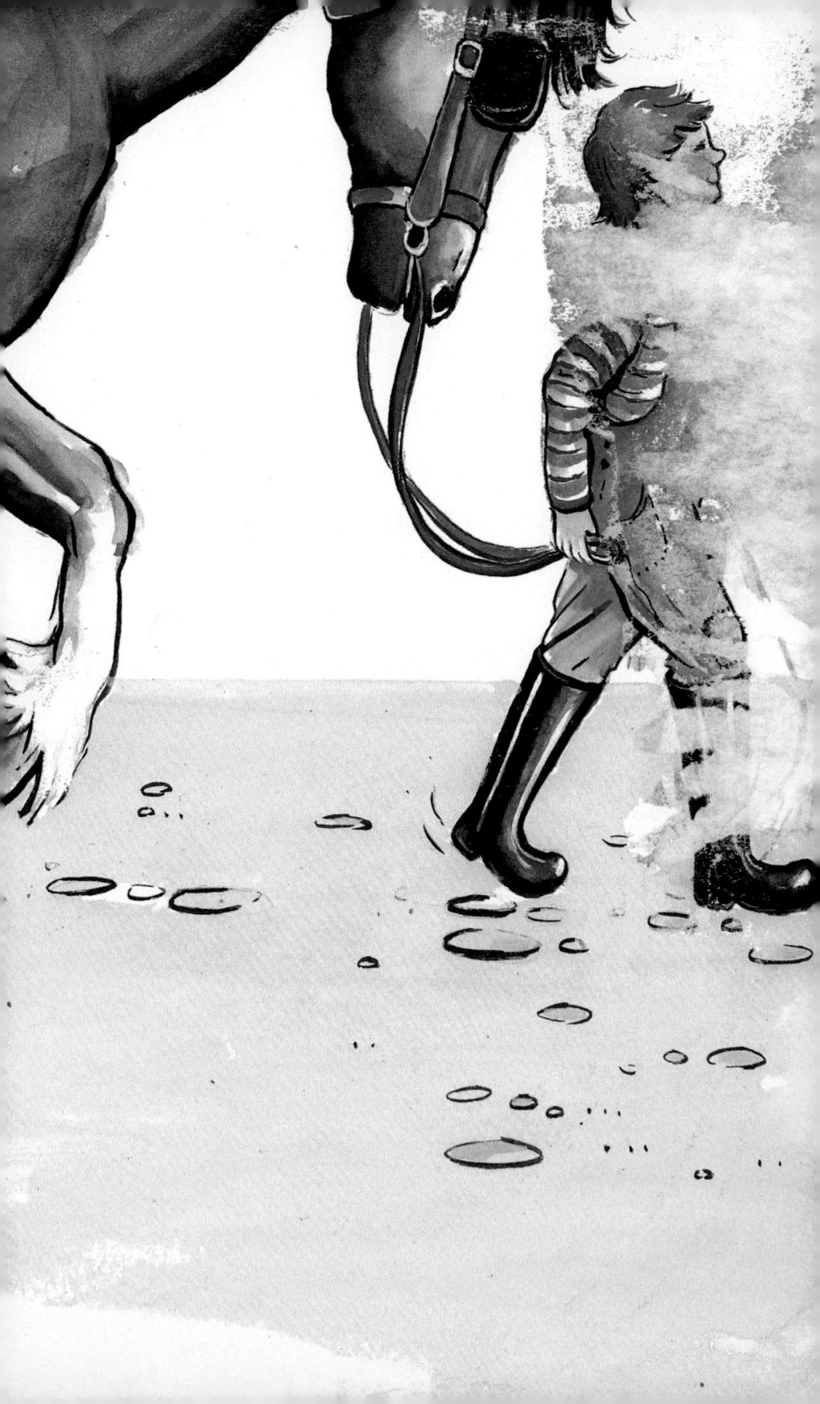

And when I grow older,
I won't only wonder,
I'll listen and learn
and start to find out;
For I'm glad that this beautiful
world which God made
Is so full of things
to wonder about.

Published by
Lion Publishing
Icknield Way, Tring, Herts, England
ISBN 0 85648 456 3
Albatross Books
PO Box 320, Sutherland, NSW 2232, Australia
ISBN 0 86760 244 9

First edition 1983

Printed in Spain by Artes Graficas, Toledo

D.L.: TO-1552-1982